Richard Katalayi Kanyinda
Ziv Kabambi Katambwe
Richard Kamangu Kalonji

Choice of building materials

Richard Katalayi Kanyinda
Ziv Kabambi Katambwe
Richard Kamangu Kalonji

Choice of building materials

on energy efficiency in buildings

ScienciaScripts

Imprint

Any brand names and product names mentioned in this book are subject to trademark, brand or patent protection and are trademarks or registered trademarks of their respective holders. The use of brand names, product names, common names, trade names, product descriptions etc. even without a particular marking in this work is in no way to be construed to mean that such names may be regarded as unrestricted in respect of trademark and brand protection legislation and could thus be used by anyone.

Cover image: www.ingimage.com

This book is a translation from the original published under ISBN 978-620-6-71260-2.

Publisher:
Sciencia Scripts
is a trademark of
Dodo Books Indian Ocean Ltd. and OmniScriptum S.R.L publishing group

120 High Road, East Finchley, London, N2 9ED, United Kingdom
Str. Armeneasca 28/1, office 1, Chisinau MD-2012, Republic of Moldova, Europe
Printed at: see last page
ISBN: 978-620-7-61892-7

CHOICE OF BUILDING MATERIALS FOR ENERGY EFFICIENCY IN BUILDINGS.

Ziv Kabambi Katambwe[1] , Richard Katalayi Kanyinda[2]

Richard Kamangu Kalonji[3]

[1]Institut du bâtiment et travaux publics (IBTP), Mbujimayi,

DR.Congo

[2]Official University of Mbujimayi (UOM), Mbujimayi,

DR.Congo

[3]Institut supérieur de commerce (ISC), Mbujimayi, DR.

Congo

ABSTRACT

In the building, the energy efficiency is related to the functionality even the building materials.The energy efficiency makes it possible to reduce the consumption of energy for a given use, while preserving the environment.The energy need in the building is not if pain-killer that, it is judicious and complex. The insulation of the external walls can bring a profit of about 22,7 %, the insulation of the roof reaches 37,3 %, whereas the insulation of the floor has a negative impact on the energy performance.

***Keywords :** Building materials, energy efficiency, thermal exchanges, simulation...*

NOMENCLATURE AND THERMAL CONCEPTS

λ   Conductivité thermique KJ/(h m K)
C   Capacité thermique KJ/(kg K)
d   Densité kg/m^3
e   Epaisseur m
E$_U$ Energie utile KWh
U   Coefficient de déperdition des fenêtres
     W/(m^2.K)
g   Coefficient de transmission des fenêtres

- Thermal conductivity (λ): is the capacity of a material to

conduct heat.

- Heat capacity (C) is the energy required to raise a body's

temperature by 1K.

- Heat loss coefficient (U): is the thermal conductivity of a

material, reflecting its ability to transmit heat by conduction.

- Thermal resistance (R): corresponds to a material's ability to

resist cold and heat.

- Useful energy (E_u): is the energy available to the consumer

after the conversion by its equipment (lights, heat, motive

power, etc.)

1. INTRODUCTION

The building sector is considered to be a high energy consumer, and this is especially true in Central Africa, where the housing stock has exploded in the space of a few decades without any consideration being given to energy management. This is how the notion of energy efficiency came to be, in the wake of a succession of energy crises.

This state of affairs means that we need to look at the most effective ways of significantly reducing the energy needs of homes, without having to invest too much money and too many people. Our work will focus first on the choice of materials and then on the appropriateness of insulating external facades.

To put the present study into perspective and identify the common points and They are part of a national and international dimension and demonstrate the growing interest in issues related to energy management, particularly in the building sector.A look at the scientific literature reveals the existence of a number of energy-saving methods. Simulation methods are very effective in building energy analyses, because they deal with most of the significant parameters relating to energy consumption.Several numerical studies have been carried out on energy efficiency and the optimisation of residential buildings.

A comparison of annual energy consumption between the different variants (roof insulation and thermal insulation) was carried out. of the walls) for each zone. The overall result is that all the proposed insulation solutions reduce energy consumption, although external insulation is a better energy solution than internal insulation. The SimulArch programme used an existing building, a T3 flat with a surface area of 75 m2. The building has a reinforced concrete structure and a fairly light breeze-block shell in the town of Mbujimayi. The measures simulated include wall insulation and glazing surface area. Simulation of the thermal behaviour of the dwelling in cold weather showed that the insulation alone reduced heat loss by around 25%. The best orientation, glazed surfaces and thermal insulation are simulated using TRNSYS software. (transirent system simulator) for a building located in the Mediterranean region, the results show an annual gain in energy consumption of around 27.59% thanks to the three measures combined and a performance energy from building of 64 KWh/m2.year.

2. HEAT EXCHANGE

Heat is exchanged in the building in four ways through the envelope: conduction, convection, radiation and evaporation or condensation.

- **Conduction**: conduction occurs when a flow of heat passes through a material by contact between the hottest molecules and the coldest molecules.

- **Convection**: Convection occurs when molecules move from one place to another, exchanging the heat they contain.

- **Radiation**: is the exchange of heat through space by electromagnetic waves.

Like light, radio waves and x-rays, but with a different wavelength.

- **Evaporation or condensation**: this phenomenon involves a change in the temperature. of liquid or gaseous state) and produces heat absorption or emission.

3. COMFORT AND ENERGY EFFICIENCY

3.1. Comfort

Thermal comfort has been defined as satisfaction with the thermal environment. established by heat exchange between the body and its environment.

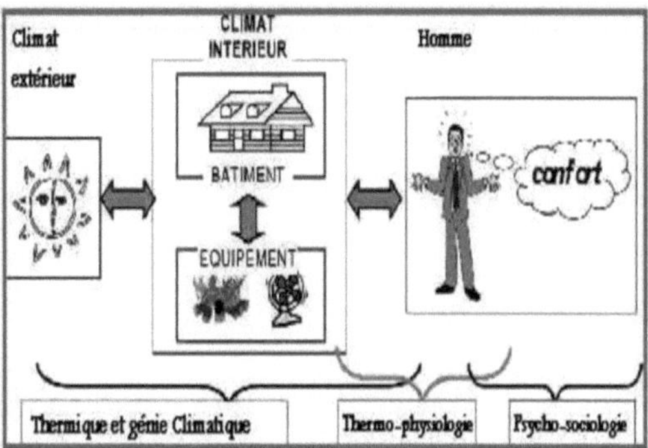

Figure 1: Phenomena involved in thermal comfort characteristics

3.2. Energy efficiency

Energy efficiency is the minimisation of energy consumption to provide a service. The design of energy-efficient buildings is a complex process, Its complexity is due to the voluminous information that requires a particular approach to the technical and architectural choices made for this type of design.

For example, the shape, compactness and orientation of a building have a significant impact on its energy performance, and the wrong choice of materials can lead to unpredictable failures that have a frightening impact on the building's energy consumption over the long term.

By embracing the energy efficiency approach, the building turns to sustainable design, and energy efficient can accommodate occupants, so energy consumption can be greatly reduced by adopting energy efficiency strategies in the building.

4. OBJECTIVE OF THE STUDY

The objective of this study is to evaluate the evolution of energy requirements (in useful energy) as a function of the choice of construction materials, their insulation and the thermal comfort conditions of a residential building located in Central Africa.

5. METHODOLOGY

In order to estimate energy requirements in this study, two dynamic thermal simulations were carried out: A base case simulation, which is based on a base model and will serve as a reference. This is based on a model of the project using the two energy efficiency measures (Choice of material and thermal insulation), each of which will result in an energy requirement that will be the point of comparison with the base case. The simulation will be carried out using TRNSYS software version 16, and its TRNBuild interface (Type 56).

5.1. Building parameters base case:

5.1.1. Geographical coordinates

The geographical coordinates correspond to the town of Mbujimayi which is located in the south-east of the Democratic Republic of Congo Latitude: 6.9° South Longitude: 23.36° East.

Average temperature: 24°C Climate: humid tropical Area:

135.12 km2 Average rainfall: 611.5 mm Population: 3367582

inhabitants Mbujimayi is located on the Kasai Plateau, a gently

undulating plateau sloping down from the west, at an altitude of

583 m. (Source: Mbujimayi town hall).

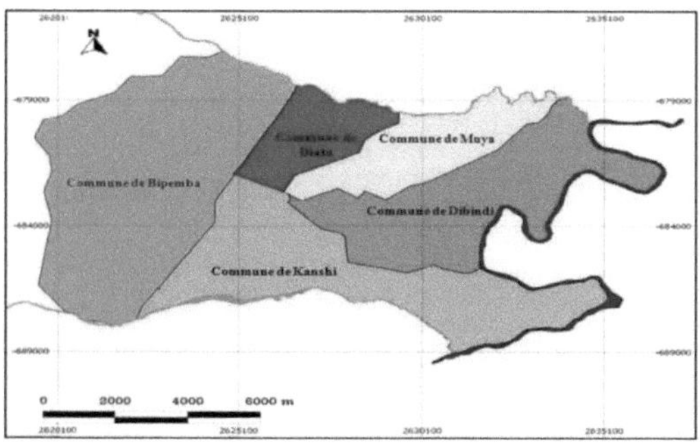

Figure 2: Map of the town of Mbujimayi

5.1.2. General plan of the building :

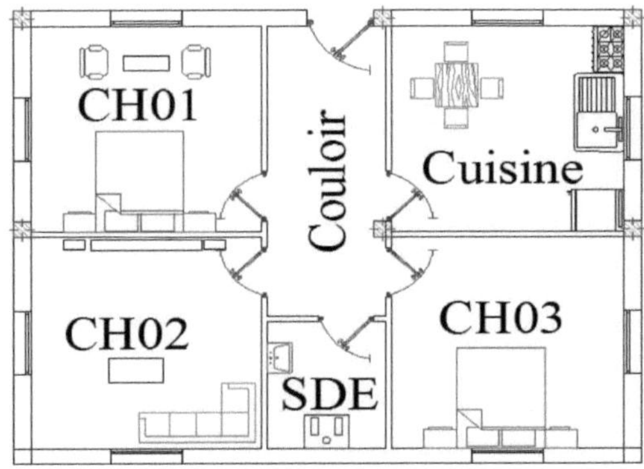

Figure 3. General plan of the base case building

5.1.3. Building dimensions base case

The base case building has a surface area of 80 m2 and a volume of 240 m3. The entrance to the building faces north. The uninsulated external walls are 15 cm thick hollow brick with cement mortar rendering on the outside and plaster on the inside; the partitions are 10 cm thick hollow brick with plaster rendering on both sides. The lower floor (the slab) consists of a

20 cm thick layer of stone followed by 10 cm of concrete, covered with tiles (the sub-floor is 2 cm thick cement mortar). The roof is 20 cm thick concrete-hourdi, with a cement mortar screed and interior plaster. The glazed area is 10% of the floor area (representing approximately 6.67% glazed area per façade). With single-glazed windows that have a U= 5.74 W/ (m2.K) coefficient and a g= 0.87 coefficient.

5.2. Type of building materials The simulation will be based on the appropriateness of the choice of materials for the external facades, so we will use five types of material in addition to our base case, namely breeze block (10 cm and 20 cm), concrete and hollow brick (15 cm and 10 cm). The simulation will include two different thicknesses for brick and breeze block. The choice of material thicknesses is based on their availability on the market. The characteristics of the materials are described in the following table:

Table 1. Characteristics of building materials

Matériaux	Conductivité thermique (KJ / hmK)	Capacité thermique (KJ / kg K)	Densité (Kg / m3)	Epaisseur (m)
Brique creuse	1.70	0.79	720	0.15
Brique creuse	1.80	0.79	720	0.10
Parpaing	4.007	0.65	1300	0.10
Parpaing	3.79	0.65	1300	0.20
Béton	7.56	0.80	2400	0.1

5.3. Impact of insulation

The impact of insulation will be at the heart of the second part

of this study, as we have opted solely for expanded polystyrene

as an insulator with the following thermal characteristics:

λ=0.141(KJ/(h m K)), C=1.38(KJ/(kg K)) and

d=25(kg/m3).

The insulation will be used in thicknesses ranging from 1 cm to

10 cm, for the external facades, the roof and the lower floor, in

order to determine both the area that needs to be insulated first and the thickness of insulation that will guarantee optimum

energy requirements.

6. RESULTS AND DISCUSSION

6.1. Base case energy requirement:

This step consists of parameterizing the TRNSYS software with the characteristic data of the base case using TRNBUILD (Type 56) and the meteorological data for Constantine, and setting the calculation step to one hour for each iteration: The energy requirement is 9180 (KWh/year) for heating and 11060 (KWh/year) for air conditioning, giving a total annual requirement of 2024 (KWh).To obtain the energy performance of our base case, we divide the total annual requirement by the surface area of the building80 m2.The energy performance of our case is of the order of 253 KWHEu/m2.year.

6.2. Energy requirements according to choice of building

materials

Table 2. Energy requirements according to choice of building
materials

Matériaux	Besoin énergétique Kwh			Performance énergétique	Economie d'énergie
	Chauffage	Climatisation	Total	Kwh / m2.an	%
Brique creuse 15cm	9180	11060	20240	253.00	
Brique creuse 10cm	10910	11590	22500	281.25	-11.17
Parpaing 10cm	12940	11780	24720	309.00	-22.13
Parpaing 20cm	10300	10690	20990	262.38	-3.71
Béton	12250	10600	22850	285.63	-12.90

6.3. Energy requirements as a function of insulation impact

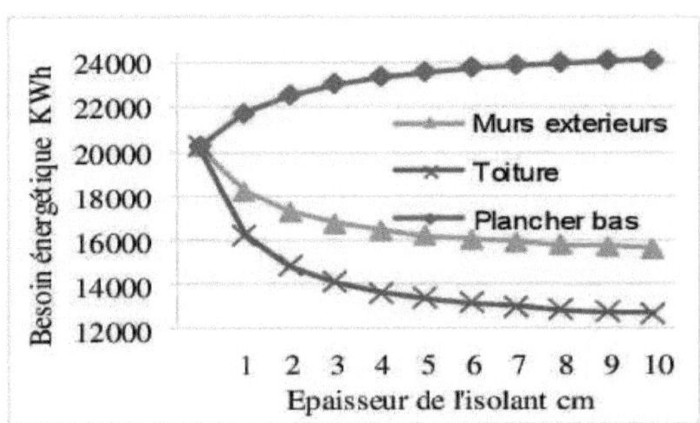

Figure. 4. Energy requirements as a function of insulation

impact

The choice of building material and the impact of insulation were important in determining the total energy requirement.

The simulation results showed that the use of breezeblocks can reduce energy performance by as much as 22.13%.

The range of impact of the choice of building materials on the useful energy requirement can be as high as 50%.

The results showed that low floor insulation had a negative impact on the total energy gain, while roof and external wall insulation had a significant impact, albeit in different proportions.

Insulating external walls can bring a saving of around 22.73%, but insulating the roof far outweighs that of external walls, by the thicker the roof insulation, the lower the simultaneous heating and cooling energy requirement, unlike external wall insulation. With regard to the thickness of the insulation, a distinction should be made between external wall insulation and roof insulation, because in the case of external wall insulation, after a thickness of 3 cm, the energy requirement curve for air-conditioning rises again, unlike the curve for roof insulation, which rises steadily for both heating and air-conditioning.

If we look at the results in detail, the first result needs to be qualified, as we can see that other parameters need to be taken into account separately: cooling and heating requirements.

Whereas the impact of insulation on energy requirements is proportional to the size of the building. the surface area to be insulated, because roof insulation in our case is the best option, and the thickness of the insulation also plays a crucial role, and a case-by-case study will help determine the ideal configuration by integrating the two properties.

7. CONCLUSION

Today, the building sector is the most important lever for optimising energy efficiency. Instead of massively subsidising the price of energy, it would be wiser to use the financial sums allocated for this purpose to finance energy efficiency measures, because just by insulating the roof the total energy gain obtained is more than 1/3 using conventional insulation, whereas this result can very well be improved by using materials that environmentally friendly materials. The energy efficiency of buildings is an essential and inescapable factor in the choice of construction materials. However, this work has shown that all factors need to be taken into consideration before any action is taken. This approach involves thinking first and foremost at the design stage to facilitate the integration of energy efficiency solutions in the most optimal way.

8. BIBLIOGRAPHICAL REFERENCES

1. Fezioui N., et all: The traditional house with horizontal

opening: a trend towards zero energy house in the hot, dry

climates, Science

2. Frank Hovorka, Guy Jover, Richard Franck, "Energy

efficiency in buildings: optimising the energy performance,

comfort and value of commercial and industrial

buildings"(2015)

3. Règlementation thermique des bâtiments d'habitation, Règles

de calcul des déperditions calorifiques Fascicule 1 (D.T.R C3-2)

1972

4. S. Foura et M. Zerouala, "Simulation des paramètres

architecturaux du confort d'hiver" Sciences & Technologie D,

vol. 1, n° %126, 2007.

5. J. Samar and A. Salman, "Optimum, technical and energy

efficiency design of residential building in Mediterranean

region," Elsevier, vol. Energy and Buildings, no. %143, 2011.

6. R. Guechchati, M. Moussaoui, A. Mezrhab and A. Mezrhab, "Simulation de l'effet de l'isolation thermique des bâtiments: Cas du centre psychopédagogique SAFAA à Oujda," Revue des Energies Renouvelables, vol. 13, n° %12, 2010.

7. Ernest NEUFEURT Les éléments des projets de construction 8emeEdition.Dunod, Paris, 2002

8. CLONED J, "Thermal insulation materials for buildings", Centre d'animation régional en matériaux avancés, May 2010.

9. COULOMB Philippe, HERMANN Guillaume, SARLIN Sophie, L'isolation thermique dans la conception et la réalisation des locaux de travail.

10. HOLLAERT, Laurie "Analysis of financial profitability and related benefits à l'isolation thermique : étude de cas adaptés au modèle belge" Master's thesis, Université libre de Bruxelles 2014.

TABLE OF CONTENTS

Printed by Books on Demand GmbH, Norderstedt / Germany